Angela Pinzón Pinto

Pedobiology

Angela Pinzón Pinto

Pedobiology

First Edition

ScienciaScripts

Imprint
Any brand names and product names mentioned in this book are subject to trademark, brand or patent protection and are trademarks or registered trademarks of their respective holders. The use of brand names, product names, common names, trade names, product descriptions etc. even without a particular marking in this work is in no way to be construed to mean that such names may be regarded as unrestricted in respect of trademark and brand protection legislation and could thus be used by anyone.

Cover image: www.ingimage.com

This book is a translation from the original published under ISBN 978-613-9-40375-2.

Publisher:
Sciencia Scripts
is a trademark of
Dodo Books Indian Ocean Ltd. and OmniScriptum S.R.L publishing group

120 High Road, East Finchley, London, N2 9ED, United Kingdom
Str. Armeneasca 28/1, office 1, Chisinau MD-2012, Republic of Moldova, Europe
Printed at: see last page
ISBN: 978-620-8-16550-5

Contents

Words can never suffice when what needs to be said overflows the soul
Julio Cortazar.

This book carries with it the influence of your teachings, and on every page, I hope to reflect the respect and gratitude I feel for you.

Dr. Abdon Cortes

Dr. Clara Chamorro

LOMBRIZ

ESCARABAJO ESTERCOLERO

ANT

TERMITA

SNAIL

CIEMPIES

PROLOGUE

Today a new path in the history of soil biology dawns: a journey through the diversity of organisms that live there with their marvellous contributions and traces that leave an emblem of their memory and the indelible trail of their work.

It is well known that Paedobiology is a rich network of connections, relationships that makes it unique, wonderful, simple and complex at the same time; in it we feel the power of life, its sharing and all the characteristics that it offers.

The edaphic fauna is the promise of a fascinating world full of adventure and uncertainty, joy and hope that works constantly in the search for balance and harmony; its representatives show how to associate with the soil, how to register the passing of time and the change of climate; with their work they show their capacity for perception, response and the connections they establish to maintain the environment in which they live.

The edaphofauna is an essential part of the landscape and is a reference of daily life, it is part of a whole, it forms an interdependent, bidirectional circuit, in a process of harmonious and constant exchange of life, united by a system that allows them to perceive and connect with each other with a level of complexity, wisdom and tacit agreements. Soil organisms are the great artists of soil life.

It is possible that with the current environmental, economic, political and social threats, limited water resources, lack of food security and the unpredictability of climate change, the importance of soil fauna will be increasingly appreciated.

The book compiles the fascinating story of the representatives of life on the ground, as they enliven the wonder of life, nurture wisdom, share the multifaceted nature of their activities and inspire solutions in the human-soil relationship, all of which benefit present and future generations.

Marfa Romero Pinto Biologist and Agrologist

PARTIE I

THE "BIG IDEA OF THE GROUND". DE DARWIN

Darwin has often been called the first soil-ecologist; in 1881, he described how earthworms produce "vegetable mould" as a result of their feeding activities and mixing of mineral and organic soil materials.

He also wrote: The plough is one of the oldest and most valuable of human inventions, but long before it existed the earth was ploughed regularly, and still is, by earthworms. I doubt if any other animals have played so important a part in the history of the world as these unorganised creatures. "Therefore I must conclude that the whole vegetable cover (soil) of the whole country has passed and will pass many more times through the intestinal tract of these worms".

He also found that the soil is formed by "worms" that engulf the mixture of dead matter and decomposed rock and expel a mixture that is the soil. He was the first to "discover" and describe how soil is formed. He was also the first to "discover" how important soil is for life on Earth. (Huhta, Veikko. (2007).

Darwin, for the first time, demonstrated the impact of ants on the genesis and modification of the soil colonised by them, in comparison with that of earthworms, since in soils where ants are abundant, they can surpass them in terms of the number of soil particles removed.

Ants, therefore, are considered second only to earthworms in bio-peat capacity, but the impact of ants can outweigh them in removal and transport.

"IS NOT THE STRONGEST OF THE SPECIES IT IS NOT THE MOST INTELLIGENT THAT SURVIVES, IT IS THE ONE THAT IS MOST ADAPTABLE TO CHANGE". DARWIN.

INTRODUCTION

Soil contains an almost infinite range of organisms that inhabit it temporarily or permanently, and fundamental processes of pedogenesis take place in it; the modification of soil characteristics by the action of soil fauna is a very important factor that must be taken into account when analysing soil genesis, morphology and properties.

The rearrangement of soil materials by biota during genesis forms organo-mineral complexes, modifies the soil surface, digs channels that transform porosity and allow water and air flow, and forms stable structures. Organisms not only contribute to the total formation of soil organic matter (SOM), but also decompose it and transform it into nutrients, e.g. C, N, P, S.

Each soil has its own range of organisms which is determined by soil conditions such as temperature, humidity, acidity, oxygen and others. The role of soil fauna in soil formation is often overlooked, sometimes we stop only at the presence of an earthworm. Edaphofauna play a vital role in shaping the soil environment by forming and modifying the soil architecture with tunnels, transporting particles and organic matter, transforming horizons, and creating new habitats in the soil.

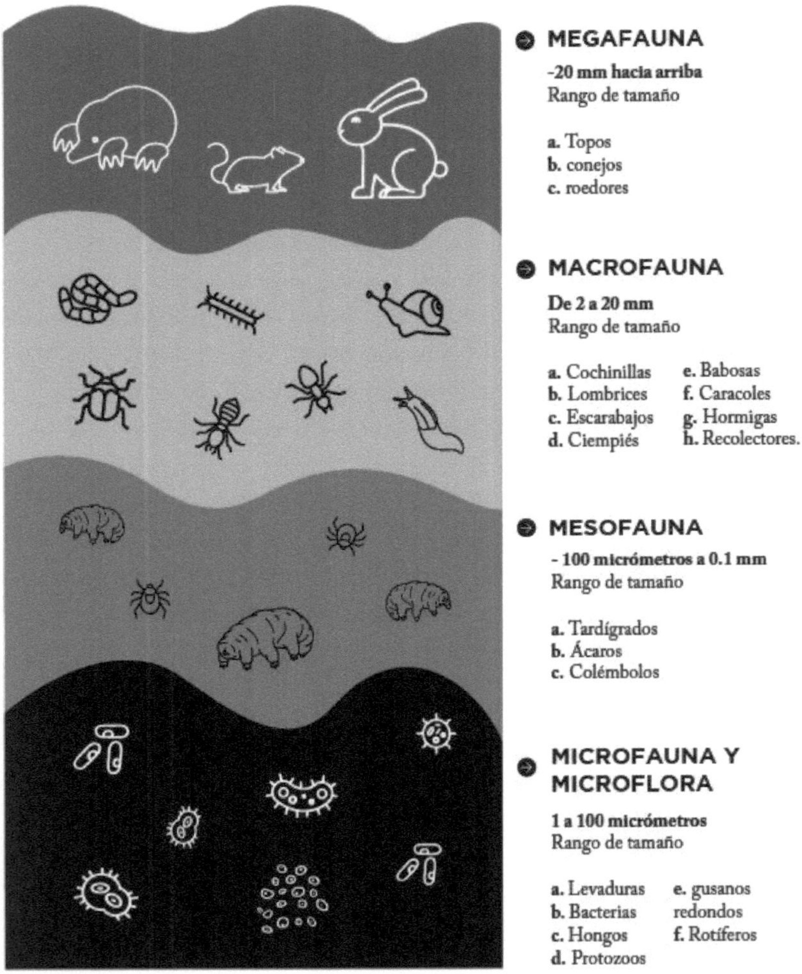

MEGAFAUNA

20 mm upwards Size range a. Moles b. rabbits c. rodents

MACROFAUNA

From 2 to 20 mm Size range a. Mealybugs b. Earthworms c. Beetles d. Centipedes

e. Slugs f. Snails g. Ants h. Collectors.

MESOFAUNA

- 100 micrometres to 0.1 mm Size range a. Tardigrades b. Mites c. Springtails Tardigrades b. Mites c. Springtails

MICROFAUNA AND MICROFLORA

1 to 100 micrometers Size range a. Yeast b. Bacteria c. Fungi d. Protozoa e. Roundworms f. Rotifers Yeasts b. Bacteria c. Fungi d. Protozoa e. Roundworms f. Rotifers

"The research on microfauna is the most researched, very little is known about megafauna, as well as mesofauna and within macrofauna the soil worm is perhaps the most studied".

Soil biota have numerous and varied functions that play an important role in determining the chemical, physical and biological properties of the soil, namely *megafauna, macrofauna, mesofauna and microfauna* (Fig. 1 ©) . In the country, research on microfauna is the most investigated, very little is known about megafauna, the same as mesofauna and within macrofauna, the soil worm is perhaps the most studied; that is why the objective of this text is to show some studies that have been carried out in the country, and also to encourage the study of soil biology. The current holistic view tends to link the diversity and functions of the soil surface with subsurface communities.

In Colombia the study of the edaphofauna began especially with the macrofauna, approximately 45 years ago in the country, and its managers were Dr. Abdon Cortes and Dr. Clara Chamorro, through the National University, Jorge Tadeo Lozano, and the Instituto Geografico Agustm Codazzi.

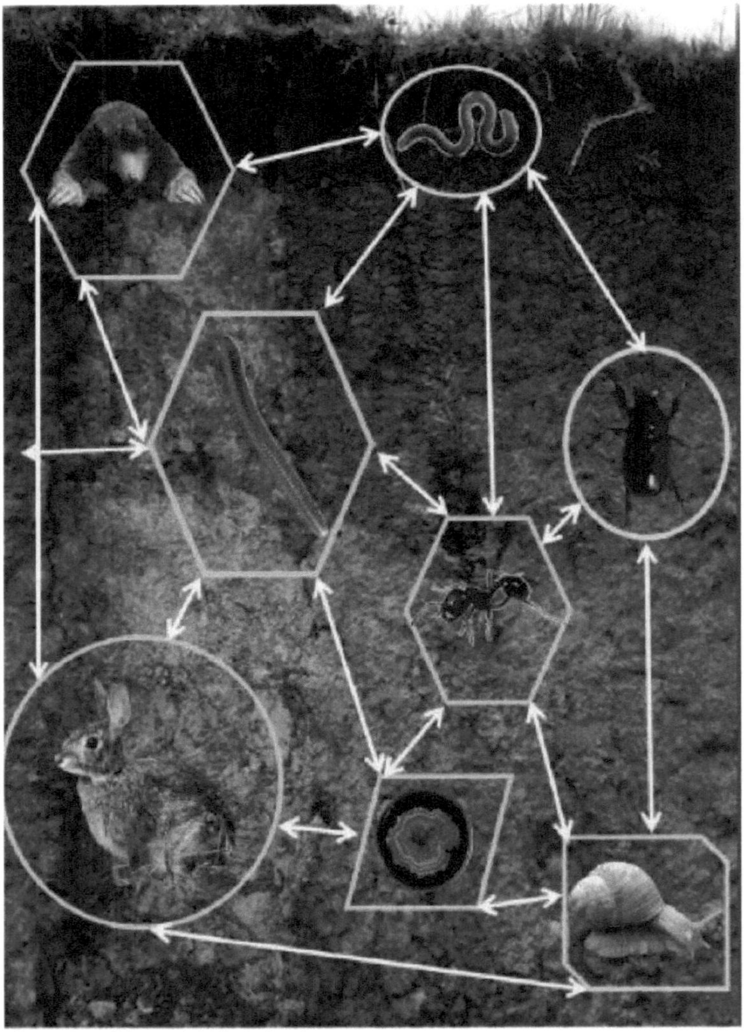

Figure 1. Recreation of the paedobiological environment.

PARTIE II

FACTORS THAT AFFECT EDAPHOFAUNA

Soil contains at least a quarter to a third of all living organisms on the planet, but only about 1% of soil organisms have been identified.

Many of the macrofaunal groups are sensitive to natural and anthropogenic environmental disturbances, which cause changes in their species composition and abundance. For this reason, they are considered to be a very important biological indicator of soil health.

The diversity and activity of soil organisms are regulated by a hierarchy of abiotic and biotic factors. Among the abiotic factors are: climate, texture, structure type, soil moisture, salinity and pH.

Climate: its influence is assessed by monitoring the physiology of soil organisms. Their activity and growth increase with high soil temperature and humidity, as this also increases their metabolism. The climatic conditions fluctuate strongly at the surface, which is why some of them migrate, either deeper into the soil profile or to other habitats that favour them.

Since climatic conditions are different in each area of the world, in each place, the components of each soil species have developed edaphotypes that present optimal ranges of temperature and humidity, and this range of adaptations depends on each specific stage of their life.

Soil texture clearly influences the activity of soil biota. For example, loamy, clayey or sandy-loam textures favour microbial growth and earthworm activity, while sandy soils with low water retention potentials are less favourable. For ants, a sandy loam texture is favourable, while for termites, a clayey texture is the most suitable.

Structure and Porosity, soil structure is associated with a complex of pore spaces that play a fundamental role in the functioning of the soil through control of

multiscale interactions of physical, chemical and biological processes.

The most favourable structure for soil biota is the granular structure, the fine and medium subangular and angular blocks harbouring meso and microfauna. The channel and burrow systems, elaborated by the macrofauna, allow aeration and oxygen dispersal in the soil and around the roots.

Salinity can increase near the surface due to poor fertilisation or drying and cause severe stress to soil organisms, however, sensitivity to salinity differs between species, and lowering salinity (by rainfall dilution) sometimes has positive effects by allowing other organisms to thrive.

Changes in soil pH: can affect the metabolism of species by affecting the activity of many enzymes and the availability of nutrients, and is therefore lethal to most organisms. Phosphorus availability is an example, as it is maximised when the soil pH is between 5.5 and 7.5.

A

LOMBRIZ

Family
LUMBRICIDAE
Class
OLIGOCHAETE ANNELIDS

Figura 2. Lombriz de suelo.

Figure 2. Soil worm.

A BIT OF HISTORY

The Sumerian civilisation, known for its great knowledge of agriculture more than 5000 years ago, established the quality of the soil from the density of earthworms when digging a hole in it. For the Egyptian civilisation the earthworms were valued to such an extent that they deified them and even punished with death those who harmed them. The Greek philosopher Aristotle in his work "Historia animalium" stated that earthworms were the intestines of the soil and contributed to its productivity.

Scientists at the Universities of Reading and York report that calcium carbonate (calcite) nodules excreted by earthworms dug from archaeological sites provide a unique measure of ancient local temperatures. Earthworm castings from the soil can be used to measure past temperatures, thus providing a window into ancient climate (Geochimica et Cosmochimica Acta.).

The knowledge of the different species of earthworms in Colombia and their activity in the soil began in the Sabana de Bogota and the surrounding paramos, then in the soils of the Department of Tolima, the Gorgona Island, the Atlantic Coast, the Coffee Zone, the Caqueteno foothills and some sectors of the Amazonian Forest and around the Migmatitic Complex of Mitu.

Earthworms are considered typical soil ecosystem engineers, providing ecological functions in agrosystems, such as: degrading organic matter, improving soil structure and porosity, (Bart et al., 2019a; Bertrand et al., 2015; Blouin et al., 2013; Palma et al., 2014). Earthworm species live in vertical galaxies that can extend several metres into the soil profile (fig. 3 -); at night they emerge to the surface to feed on decomposing leaves and leaf litter.

DISTRIBUTION OF THE SOIL WORMS

Between 70 and 92 % of the earthworms inhabit the horizons located in the first 20 centimetres of depth (O, A, Ap) and only 20 % or less inhabit the subsurface horizons (AB and B) between 20 and 50 centimetres of depth, (figure 4 O). It is important to clarify that earthworm populations and their activity cannot be explained as a function of a single soil property, but as the result of the interaction of internal and external characteristics of this ecosystem component.

"The knowledge of the different species of earthworms in Colombia and their activity on the ground started in the Savanna of Bogota and the surrounding paramos".

Figura 3. Canal elaborado por la lombriz del suelo (Foto A. Pinzón)

Figure 3. Channel made by the earthworm (Photo A. Pinzon)

Figure 4. Earthworms in the first soil horizons (Internet photo. Ecomandanga-org)

The analysis of the distribution of earthworms in the soil allows us to establish that both the height above sea level and the depth of the different horizons of the soil profile and their properties influence the location of these macro-organisms. According to Chamorro (1990), the higher the altitude above sea level and the greater the depth, the lower the populations of these macro-organisms. According to Chamorro (1990), the higher the altitudes and depths, the lower the earthworm populations; in the flat part of the Sabana of Bogota, the same author found differences in earthworm populations in three contrasting soils of the Cota (Hapludands), Rfo Bogota (Fluvaquents) and Techo (Haplustalfs) series; the first of them hosts 77 % of the studied population, the second one supports 20 % and the last one 5 %.

"

IT IS IMPORTANT TO KNOW WHAT HAS HAPPENED WITH CLIMATE CHANGE, HOW THE HABITS OF THESE ORGANISMS HAVE CHANGED AND WHAT THE REPERCUSSIONS WILL BE ON SOILS AND WHAT CONSEQUENCES WILL BE SEEN AT THE LEVEL OF TERRESTRIAL ECOSYSTEMS; THIS IS A GREAT TASK FOR THE NEXT GENERATIONS OF RESEARCHERS.

BIOGENIC AGGREGATES

The biogenic aggregates or moulds expelled by earthworms are produced naturally as part of their life cycle, after they have absorbed all the nutrients, they excrete their nutrient-rich faecal matter, which is used by the plant rafc.es, (Fig. 5 O).

A study by Emma Versteegh, comments: "Earthworms secrete several pellets a day, so they provide a record of seasonal variations in temperature, as well as local geographic variations". Scientists at the Universities of Reading and York (2013), found that calcium carbonate (calci- ta) droppings dug from archaeological sites provide a unique measure of ancient local temperatures.

Figure 5. Worm droppings on the surface (Taken from the internet Photo: Country life.)

In most soils, earthworms play a key role in the formation of aggregates and pores (figs. 6● and 7●) in the soil. These biogenic aggregates can represent more than 50 % of the soil volume. These formations are dynamic and can change in a short period of time, therefore, soil management can greatly affect the biota, which means that proper management is required to sustain the organisms and their microhabitats. Figure 8● shows the droppings in a CT scan.

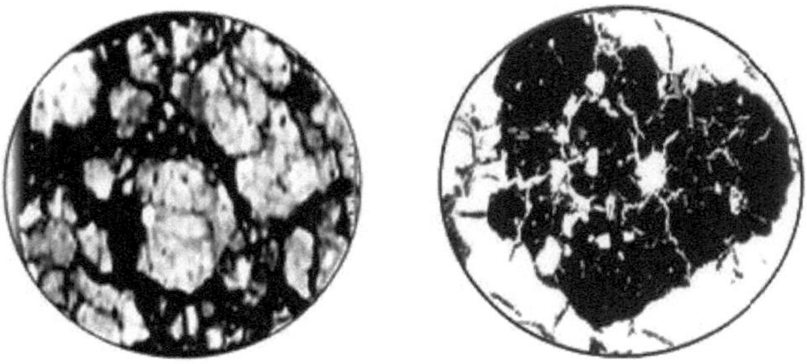

Figure 6. Computed tomography.

Soil porosity obtained from aggregates and biogenic pores (A. Pinzon) steromicroscope (A. Pinzon)

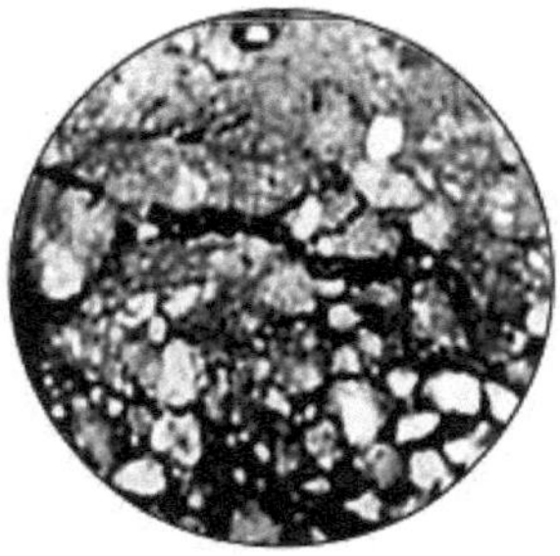

Figure 8. Computed tomography of earthworm excreta (A. Pinzon)

In three representative soils of the mapping units Mitu Complex and Tipiaca Association (Vaupes, Colombia) the earthworm populations were analysed and it was established that Eisenia constituted 50% in the soil classified as Petroferric Acroperox; Andiodrilus 39% in the Haploperox soil and 11% in the Quartzipsamments (Soto, 1986).

SOIL WORMS AS PROTEIN PRODUCERS AND PROCESSORS OF BIODEGRADABLE WASTES

Soto, 1986, analysed the protein content of Andiodrilus bogotaensis, Andiodrilus sp., Eisenia fetida, Eisenia m, Glossoscolex sp. and Glossoscolex giganteus, and also evaluated the quality of the compost produced by some of them, from the transformation of biodegradable residues such as bovine manure, coffee cherry and fruit residues.

A. bogotaensis decreased the acidity of the Cryumbrepts soil of the Chisaca series from 4.2 to 5.2 in the compost from the diet with fruit residues, which shows that the activity of the earthworms contributes to making the conditions of the environment they inhabit less drastic.

BROMATOLOGICAL ANALYSIS

The analysis carried out by Soto revealed the contents of protein, fibre, ash, fat, non-nitrogenous elements, phosphorus and calcium of the earthworms E. fetida and Glossoscolex sp. The highest values correspond to those of the protein (between 54.77% and 68.17%), which is why aminograms were carried out for this fraction. The value found in the earthworms of the Mitu soils was 51.9% (wet basis) in Eisenia sp., in the A horizon of the soil classified as Petroferric Acroperox, in an indigenous farm. The analysis of the quality of Eisenia fetida and Glossoscolex sp. proteins reported 19 of the 20 known amino acids.

All these analyses will contribute to reducing fertiliser requirements, thus reducing production costs and the ecological impact on the ecosystem through the elimination of pollutants; it will also allow the recycling of biodegradable materials, the improvement and conservation of the physical, chemical and ecological properties of the soil.

EARTHWORMS AS BIO-INDICATORS OF SOIL CONTAMINATION

Romero (1986) established that Eisenia fetida accumulates Chromium, Lead and Zinc in a higher proportion in the faeces than in the tissues, while Cadmium accumulates more in the tissues than in the faeces. It is noteworthy that the eggs of the worms studied are more susceptible to the accumulation of Cd, Cu and Pb than the hatched organisms. The results of the study show that Eisenia fetida can be used as a bioindicator of soil contamination by heavy metals and A. bogotaensis as a transformer of these into beneficial elements for plant production.

Research done by Pierre G. (2023) revealed that vermicompost alters the physical and chemical properties of animal manure faster than conventional composting and produces optimal macronutrient values, in C-Organic.

Earthworms and plants can decrease greenhouse gas emissions from soil by modulating fluctuations in soil moisture and soil macroporosity, (Pierre Ganault (2023). Furthermore, by promoting the incorporation of organic carbon into the soil, earthworm activity can decrease greenhouse gas emissions, contributing to the regulation of the Earth's climate.

Therefore, knowing the different earthworm species, their capacity to transform soil materials and the optimal conditions for the development of their activities will allow the soil scientist to recommend management practices to improve the soil, through the increase of certain species.

B

ESCARABAJO ESTERCOLERO

Family

SCARABEIDA

Class

INSECTA

Figure 9 Dung beetle

Beetles with more than 380000 described species occupy almost all terrestrial habitats and represent about 40% of the insect diversity and 25% of the total known species on the planet (A. Pinzon. 1987).

Scarabaeidae is one of the most numerous families and includes species with diverse feeding strategies; the feeding of adults in many cases differs from that of larvae, these species can be grouped into two broad categories: phytophagous, those feeding on plant matter, and saprophagous, feeding on decaying organic matter.

Scarabaeidae contribute important ecological functions such as nutrient recycling through incorporation of organic matter into the soil, soil aeration, secondary seed dispersal as a result of burial of empty excrement containing seeds. At the same time, these ecological functions provide beneficial ecosystem services mainly for the livestock industry. There is a marked difference between the resources that serve as food for the adults and the resources that are actually used as provision for the brood balls, which are undoubtedly the food for the larvae (Halffter and 2009).

The studies carried out by Pinzon on the dung beetle, family Scarabeidae, in the soil of the Bacata series, classified as Dystrand- epts located in the municipality of Madrid and the Tropaquepts andico soil, located in the municipality of Mosquera de la Sabana de Bogota, showed that this organism is very demanding with regard to soil moisture; when moisture conditions are scarce the beetle goes deep into the profile.

The activity of dung beetles is reflected in the soil in three ways: by opening channels and tunnels and by forming dung balls called crotovines Figures: 10 -11 © and 12-13 ©.

Figure 10 and 11. Material removed by the beetles (Photos A. Pinzon)

Figure 13. Crotovina (A. Pinzon)
Figure 12. Transposition of crotovin into the tunnel (Taken from internet)

For the production of crotovines, the beetle builds a ball with organic matter and manure, rolls it with its hind legs to where it has made a tunnel or hole, and deposits the eggs inside, where the larvae or chizas will develop. In order to find out how deep the beetle goes in order to leave the crotovina, the author made a material with plaster and introduced it into different tunnels and found that the depth and shape of the tunnels depend on the physical characteristics of the soil, figure 14 O.

"For the elaboration of crotovins, the beetle builds a ball with organic matter and manure, rolls it with its hind legs to where it made the tunnel or hole, deposits the eggs inside, where the larvae or chiza will develop".

O **Figure 14.** Excavation depth of tunnels (A. Pinzon)

The concentration of nutrients in the digestive tract of the chiza, according to laboratory analyses carried out by Pinzon, indicates that these organisms contribute to the enrichment of the soil environment with nutrients such as phosphorus and calcium, as well as the decomposition and mineralisation of organic matter; in addition to this process there is the effect of the excrement, its activity in transporting organic materials deep into the soil and the contribution of the exuviae that the organism leaves behind during its metamorphosis.

"

MUCH REMAINS TO BE RESOLVED ABOUT THE IMPORTANCE OF THE MANY SPECIES THAT LIVE IN THE SOIL, IN TERMS OF THEIR FUNCTION AND BEHAVIOUR.

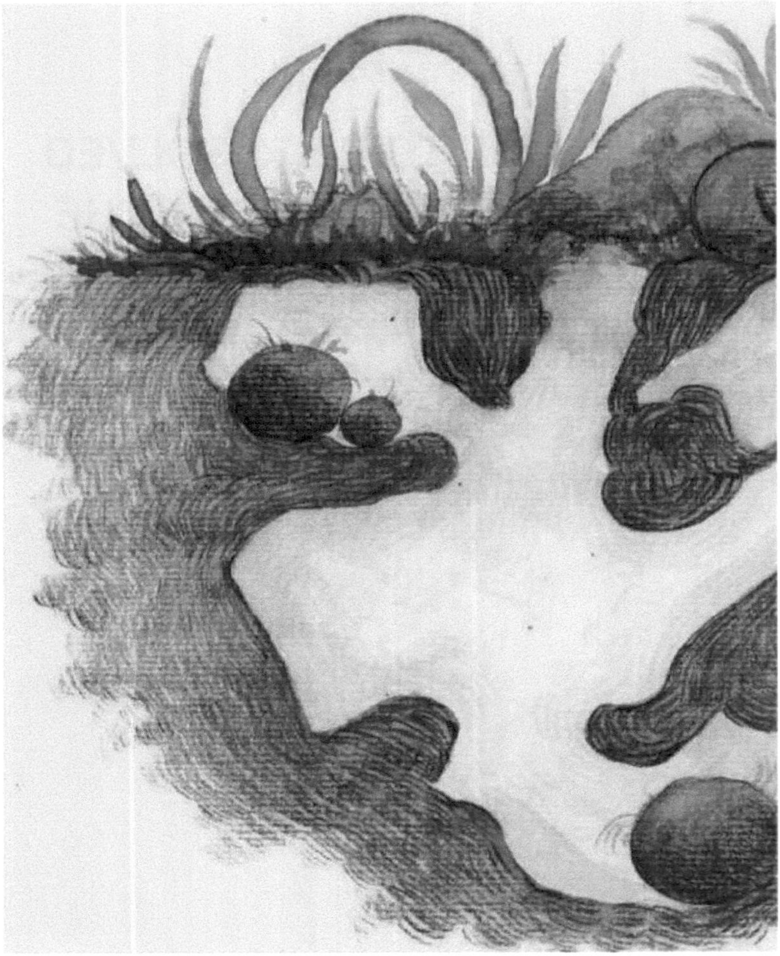

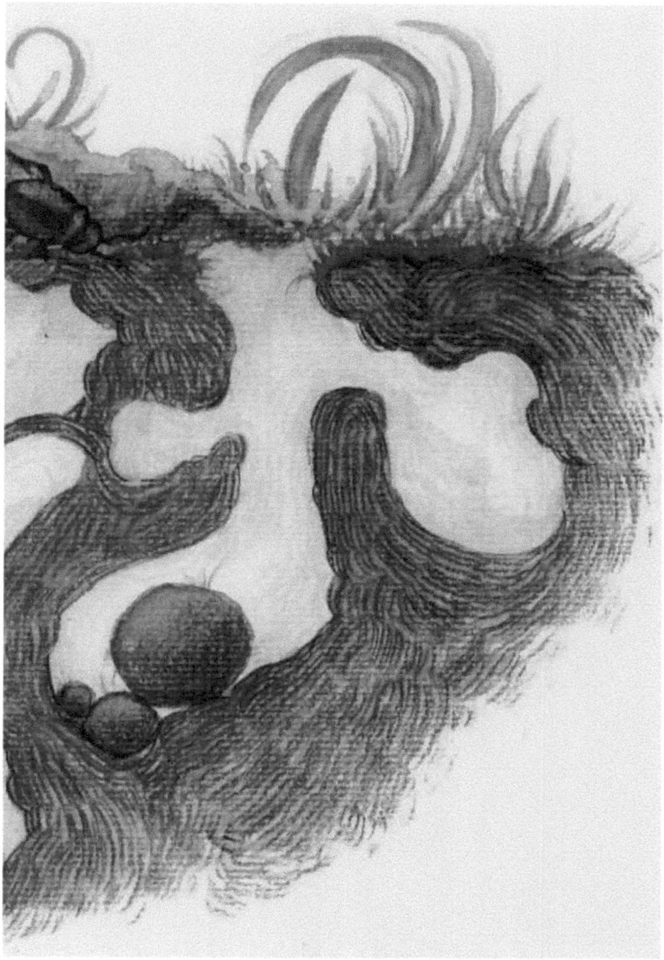

Figure 15.
Excavation of tunnels and deposition of crotovina.
Illustration Luna Pinzon

C

ANT

Family

FORMICIDAE

Class

INSECTA

O Figure 16. Ant. Taken from the internet.

These insects are great transporters of waste, such as pieces of leaves, dead insects, etc., which are taken to the anthills to feed their larvae (Figure 17). In this way, they incorporate organic matter into the soil and help it to decompose and transform it into nutrients for the plants. Another service offered by ants is soil aeration, improving the hydraulic properties of the soils in which they build their nests.

ANTS AS SOIL BIOTURBANTS

Ants play a key role in the regeneration of eco-systems. They are mainly found in sandy and sandy-gravelly soils, in which they transport this material to make their niches (Figures 18 and 19).

Figure 17. Galenas formed by the ants. Illustration Luna Pinzon

Changes in vegetation composition, coupled with physical and chemical degradation of the soil, can impact on species abundance and composition (Chamorro et al. 1991) and affect the basic properties or attributes of ecosystems and the goods and services they provide to humankind. Ants are also bioindicators of disturbance because of their high diversity and abundance, the variety of niches they occupy, their rapid response to environmental changes and their relatively easy identification, they can be useful in assessing biotic responses to agricultural practices such as fertilisation, spraying and burning.

Figure 18 and 19. Nests or habitats of the ants (Photo A. Pinzon)

"

IT IS HIGH TIME TO RECOGNISE THAT SOIL BIODIVERSITY PROVIDES MANY ESSENTIAL ECOSYSTEM SERVICES FOR HUMANS AND THE ENVIRONMENT.

ANT-INDUCED SOIL MODIFICATION AND ITS EFFECT ON BELOW-GROUND BIOMASS OF PLANTS

Along with the modification of soil chemistry, changes in soil physical properties and in the vertical distribution of nutrients are recorded. The ant nests were characterised by a predominance of sand particles and a lower bulk density (Pinzon, 1990).

The workers feed on the sap they take directly from the leaves and stems (figure 20 ©) of different types of plants and plant debris, which they mix with saliva to form the substrate for the fungus culture (figure 21 -), which they associate in mutualistic symbiosis with crumbs inside the anthill chambers. The cultivated fungi are used to feed the larvae, using bacteria of the genera Pseudonocardia and Streptomyces, which they use as pawns for the production of metabolites that protect their precious cultures.

In summary, the construction of the nests or chambers made by the ants modifies the physico-chemical properties of the soil, and this is manifested in an acceleration of nutrient cycling caused by the incorporation of organic matter.

Cultured fungi are used to feed the larvae, using bacteria of the genus Pseudonocardia and Streptomyces as pawns for the production of metabolites that protect their precious cultures.

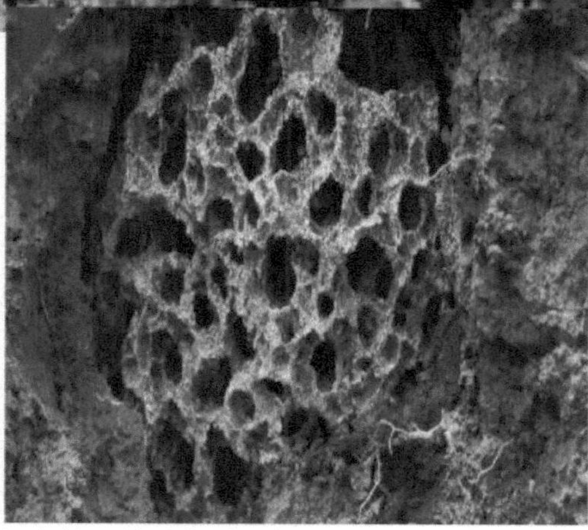

Figure 21. Mushroom cultivation inside a niche (taken from internet)
Figure 20. Leaf-cutting ants.

D

TERMITA

Family

TERMITIDAE

Class

INSECTA

Figure 22. Termite. Taken from the internet.

Termites are social insects that form colonies with different castes or groups of individuals. The castes differ both in their appearance and in the function they perform within the colony. Subterranean termites are organised into three different castes: breeders, workers and soldiers.

The termites live without leaving this particular skyscraper: the workers constantly bring food to the queen's central chamber and transport the eggs to nearby nurseries. Although temperatures fluctuate inside, the humidity is crucial for the colony's fungi (fig.23 ®) and is always maintained at 80%. A truly prodigious architectural and living structure.

Galvis, Chamorro and Cortes, during their fieldwork, observed monticules in the landscape (Fig. 24 O), with an average height of 50 cm, whose internal architecture was characterised by the presence of a

Q **Figure 23.** Termites cultivating their fungus (photo from internet)

complicated system of chambers, channels and tunnels connected by small openings. The shape of the termite mounds means that they can be suitable microhabitats not only for the termite mounds, but also for other groups of invasive organisms such as termites, beetles, beetles and spiders.
Andrea Perna, in his book "The Art of Building a Nest" in order to know the architecture and symmetry of termite mounds with an astonishing architecture and regularity, (figures 25 ©). Each construction worker deposits soil pellets in places where they are already abundant; the co-ordination of activities arises from these individual actions through a process of self-organisation.
Observations by the authors determined that termites are the result of the orderly and continuous accumulation of dejections, which are the end product of the ingestion of soil materials and plant debris transformed by termites in their digestive tract and subsequently excreted as a dark-coloured paste.
Figure 24. Termite mound (photo A. Pinzon)

Figura 24. Termitero (foto A. Pinzón)

Figura 25. Interior de un nido de termitas reconstruido mediante tomografía de rayos X. (Fuente: Andrea Perna)

Figure 25. Interior of a termite nest reconstructed by X-ray tomography (Source: Andrea Perna).

EFFECTS OF TERMITE ACTIVITY IN SOILS

The morphological and physico-chemical studies carried out on the soils of the study area show that termite activity alters the morphology and the physical and chemical properties of the soil through the transport of materials, the construction of their nests and the digestion of organic and mineral elements which they then deposit in their structures.

ALTERATION OF THE soil morphology

Termite activity alters the micro relief of the terrain by the formation of nests in the form of mounds (Figure 26 -) which persist on the surface even after they have been abandoned and subjected to erosion processes. Another effect of termites affecting the morphology of the soil profile is the construction of tunnels and galenae in some horizons and the filling of these cavities with materials translocated from the surface horizons.

Q **Figure 26.** Landscape alteration (Photo A, Pinzon)

”

SOIL IS THE CRITICAL AND DYNAMIC REGULATORY CENTRE OF MOST ECOSYSTEM PROCESSES.

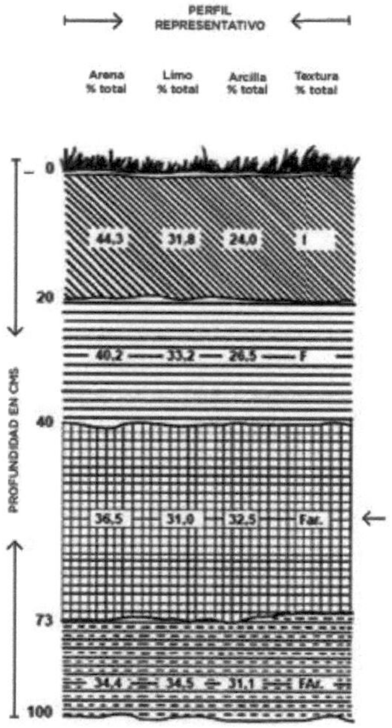

Sand Silt Clay Texture Texture % total % total % total % total % total % total

CHANGES IN PHYSICAL PROPERTIES

Galvis found higher clay content in the epigeal zone of the termites compared to the basal zone and the underlying surface soil can be considered as an indication that the termite studied select, to a certain extent, the size of the particles they carry for nest construction. The results of the determination of the particle size distribution (fig.27 ©) suggest that the termites in the studied area obtain soil material for their structures from the deep horizons.

For nest construction, termites select mainly the clay fraction and transport and transform an average of three pounds of soil per square metre (Galvis et al); this activity alters the micro-relief and morphology of the soil profile. These mounds are built with a mixture of soil, manure and termite saliva. In fact, there are also underground nests (hypo- geos), above ground (epigeos). Moisture retention, as well as pH, organic matter, ion exchange capacity, usable phosphorus and total bases of termite mound materials are higher in the epigeous zone than in the base and surrounding soil. They also undergo significant changes in bulk density and pore space. Similar data are found in the Sabana de San Benito Abad in the department of Sucre (Table 1 ©).

Figure 27. Comparison of soil particles and those found in the termite mound (Galvis).

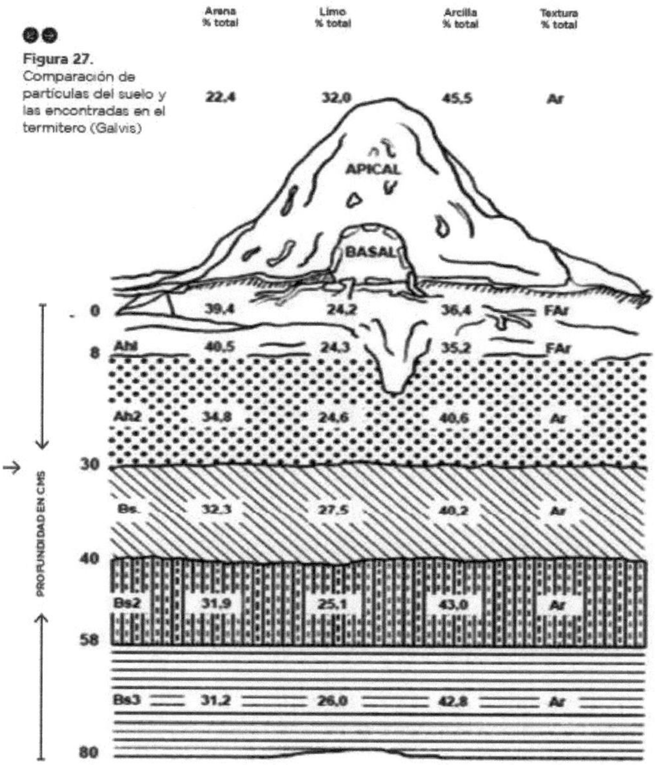

SAMPLE	pH	CALCIUM me/100g	MAGNESIUM me/100g	PHOSPHORUS p.p.m.	CARBON % % CARBON % CARBON % CARBON % CARBON % CARBON % CARBON
SURROUNDING GROUND	5.2	0.1	0.1	2.0	0.47
THERMITER	5.7	4.4	0.8	6.3	4.92

E

SNAIL

Family

HELICIDAE

O **Figure 28.** Snail, (Photo: Angela Pinzon)

Pinzon and Romero (2003), investigated some aspects related to Helix aspersa, today known as Cryptomphalus aspersus, in the Haplustalfs soil in Cucunuba, Cundinamarca, (fig.28 ®). Since snails and slugs are common, many aspects of their ecology are unknown; most of the food of these animals is fresh plant material containing cellulose and hemicellulose.

Snails in general need calcareous soils; humidity is essential for their biology as it regulates their activity. The snail requires a relative humidity of 75 to 90%, and an increase or decrease in humidity can affect its vital functions. The suitable temperature range is between 16 and 24°C and, like humidity, much higher or lower temperatures paralyse their activity.

To lay its eggs, the snail digs a hole with the front part of its foot, about 3 to 4 cm deep and 3 cm in diameter, and once the hole is made, it lays the eggs (Figure 29). Incubation lasts from 15 to 20 days depending on environmental factors.

O **Figure 29.** Snails on Haplustalfs soil (photo A. Pinzon).

(FAO, 1988). Once hatched, the juvenile remains for 5 to 10 days in the hatching chamber feeding on egg structures.

Snails are organisms that, like all soil-dwelling organisms, cause both physical and chemical transformations due to the mucoproteins and celluloses found in their faeces. These mucoproteins (figure 31 ©), help soil particles to bind together and form stable aggregates, as demonstrated in this research.

The locomotion of this organism is very slow, it was observed that it took 1 hour and 15 minutes to move 1 metre.

Snail slime is composed of proteins, glycolic acid and elastin; it also contains hyaluronic acid, copper peptides, antimicrobial peptides and other elements such as copper, zinc and iron and a pH between 6.5 and 7.6.

Figura 30. Huevos del caracol (Foto A, Pinzón)

Figura 31. Mucosidades del caracol (Foto A, Pinzón)

Figure 30. Snail eggs (Photo A, Pinzon)
Figure 31. Snail mucous membranes (Photo A, Pinzon)

"Snails are organisms which, like all soil-dwelling organisms, cause both physical and chemical transformations due to the mucoproteins and celluloses found in their faeces.

These gastropods contribute to the recycling of organic matter, even though their consumption is not very high, they promote the work of soil micro-organisms. In addition, the fragmented litter provides a source of food for other animals which can further stimulate microbial activity.

The data obtained in this research allow us to conclude the following: their faeces are an ideal medium for the activity of soil microorganisms, both the faeces and the slime they secrete during their movements contribute to the structure of soil particles, as well as to the addition of nutrients to the soil medium due to the contribution of mucoproteins. By mixing organic and inorganic fractions, they alter the soil structure, porosity, water retention, gas exchange and the chemical and nutritional properties of the soil, helping to regulate mineralisation and humification; they also increase the content of phosphorus, magnesium and potassium and, to a greater extent, calcium (Romero and Pinzon).

SNAILS AS BIO-INDICATORS OF CONTAMINATION OF METAL ELEMENTS

It is very common for soils to contain traces of heavy metals such as copper, lead, zinc, nickel, cadmium and mercury, which can reach toxic levels for plants and animals. There are species of snails such as Cryptomphalus aspersus (Helix aspersa) and slugs that store these metals in their digestive gland or in different tissues of their body. By analysing their organs we can get an idea of the levels of metals in different areas. It has been shown that they can accumulate higher concentrations than other invertebrates (Rodriguez T.).

"

SOIL ORGANISMS CONTRIBUTE TO A WIDE RANGE OF ECOSYSTEM SERVICES THAT ARE ESSENTIAL FOR THE FUNCTIONING OF NATURAL AND MANAGED ECOSYSTEMS.

F

CIEMPIES

Family

MIRIAPODOS

Class

CHILOPODS

ф **Figure 32.** Ciempies. Internet tornado.

They play an important ecological role in controlling populations of other arthropods and participate in aeration and nutrient cycling in the soil. They are vegetarians, feeding on fresh or decomposing organic matter, preferring half-decomposed leaf litter because they are unable to digest cellulose. They therefore have to ingest a very large volume of plant debris to extract it. These or- ganisms have a great diversity of habitats, they can be found under stones, in leaf litter, on the ground and under the ex- crements of large herbivorous mammals (Riojas et al., 1969). Their distribution and activities depend, to a large extent, on the shape of their bodies and on moisture relations.

They need a sufficient quantity of sugars to satisfy their needs, hence they are very important mechanical grinders and mixers.

Pinzon collected the droppings left on the surface (fig. 33 ®) in a soil under palm cultivation; the results of the chemical analysis of the droppings show phosphorus, calcium and organic carbon contents higher than those found in the soil.

The body musculature of centipedes enables them to make holes several inches deep in the soil and they rise and fall in the soil in response to seasonal changes, contributing to pore formation and water and nutrient transport.

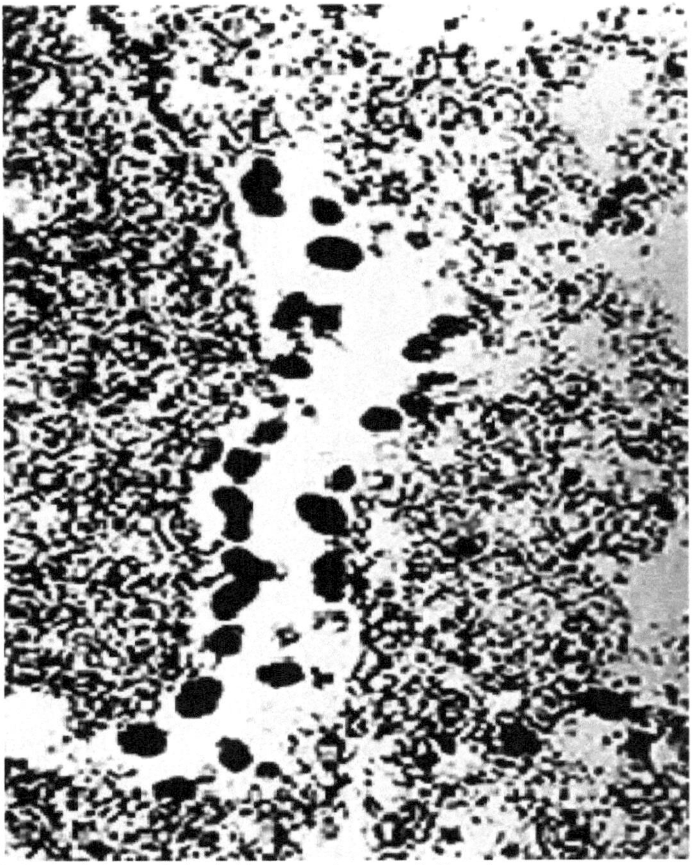

Q **Figure 33.** Centipede faeces obtained with a stereomicroscope (A. Pinzon).

The results indicate that these organisms, as representatives of the myriapods, consume large quantities of leaves of little nutritional value and excrete most of them relatively unchanged physically, but highly fragmented, and thus easily exploitable by micro-organisms.

According to Lavelle et al. (1994) the consumption of litter by these birds results in disintegration, some hydrolysis of cellulose, hemicellulose and pectin and the activation of micro-organisms in the faeces, which can be retained after days or weeks. Some consume a certain amount of soil, but do not excrete a mull humus but a mulliform moder. Among both milfoil and centipede, there are some species that mine and others that feed and live on the surface.

EDAPHOFAUNA IN A HIGH ANDEAN FOREST OF THE MONSERRATE REGION

Chamorro and Garda, studied the edaphofaunal composition of the epigeous levels, i.e. in the organic horizon and in the hypogeous horizon, up to depths of 40 cm. in physiographically characteristic zones of the paramo of Monserrate (3.300 m.a.s.l.), located to the east of the Sabana of Bogota and classified as: Dystrudepts and Humitropepts.

The analysis of the data allowed us to know the values of abundance, frequency, constancy, dominance and biomass; in addition, the biological parameters were correlated with climatic and edaphic parameters, as these are directly responsible for the maintenance of the ecosystem.

The epigean animals that inhabit part of the organic profile belong mainly to the classes Insect, Arachnida and Oligochaeta and make up 88% of the total analysed. Ninety-two per cent of the hypogean d^pteran and Coleoptera organisms are preferentially located in the first 20 cm of the mineral soil; the remaining 8% are located in the second depth (20-40 cm) and are generally lumbricids.

CHANGES IN THE MACROFAUNAL COMMUNITY IN SOILS OF VICHADA (QUARTZIPSAMMENTS) AFFECTED BY BURNING

CHAMORRO and AMAT, investigated the incidence of burning in Vichada (Colombia) on the edaphofaun- mation community in experimental plots in the Marandua region. The classes Insecta, Arachnida and Oligochaeta were studied during the 16 months of experimentation and under the 3 treatments applied (no burn, burn and recurrent burn). The number of individuals decreased by 55% as a consequence of the first burn, but recovered and increased in variety during the 10 months following the first burn.

TECHNOLOGICAL ADAPTATION OF VERMICULTURE IN THE COFFEE-GROWING AREA OF ALBAN CUNDINAMARCA

Vermicompost production with Eisenia fetida and Glososcolex sp. was tested under field, greenhouse and laboratory conditions. The worms were fed with cattle manure, coffee cherry and biodegradable waste alone and in combination. The results shown by Chamorro indicated that diets in which coffee cherry was included produced the best effects on animal biomass yield and on soil physicochemical properties such as CCC, total base content, organic matter, total nitrogen, phosphorus and microelements. The activity of Glossoscolex sp. on the above properties is slightly lower than that of E. fetida, which can be mass-produced under semi-controlled environmental conditions. The use of vermicompost can save about 50% compared to traditional fertilisation.

"

SOIL IS HOME TO APPROXIMATELY 25% OF THE PLANET'S BIODIVERSITY AND, ACCORDING TO THE FAO, DIRECTLY OR INDIRECTLY PROVIDES 95% OF THE FOOD WE EAT.

BIOEDAPHOLOGICAL STUDY PRELIMINARY OF THE ISLAND OF GORGONA

Soil fauna play an essential role in the formation, evolution and development of the substrate they inhabit. However, this is a reciprocal interaction, as the conditions of the soil environment in turn determine the biological processes occurring within it. The edaphofauna together with the vegetation, active factors in soil formation, interrelate with climate, relief, weather and parent material, thus determining the different soil classes.

The soils of Gorgona Island are mostly lnceptisols, characterised by an exuberant vegetation cover over a large part of their surface and by being subjected to tropical conditions of high humidity and high temperatures during most of the year, as a consequence, the organic matter in the process of decomposition is incorporated into the soil cycle. The results of the sampling carried out by Chamorro corresponding to the nine profiles, eight classes were recorded: Insecta. Arachnida. Diplopoda, Quilopoda. Symphila, Crustacea, Clitelata and Gasteropoda, belonging to the phylum: Artrhopoda. Annelida and Mollusca.

ARTHROPOFAUNA ASSOCIATED WITH HIGH ANDEAN FOREST SOILS OF THE COLOMBIAN MASSIF

Chamorro and Camero identified soil fauna in two high Andean forests of the Colombian Massif to evaluate differences in composition, abundance and diversity with respect to grassland areas generated by deforestation of the natural cover.

Differences were found in the composition, abundance and diversity of soil fauna depending on the type of vegetation cover, with greater similarity between the fauna of forests in different areas than between forests and grasslands in the same locality, thus inferring the influence on the biological structure at landscape scales. Biological diversity decreases with soil depth regardless of vegetation cover. The influence of deforestation of natural forests on faunal diversity is evident.

ECOLOGICAL CHARACTERISATION OF NATIVE EARTHWORMS (PHERETIMA SP., EUDOVOSCOLEX SP. AND PERISCOLEX SP.) UNDER DIFFERENT LAND USES (GUAVIARE, COLOMBIA)

Chamorro and Leon characterized the earthworm populations of Pheretima sp., Eudovoscolex sp. and Periscolex sp. in a soil classified as Typic Dystropepts, in the municipality of El Retor- no (Guaviare, Colombia), under eight types of use: ungrazed grassland, semi-grazed grassland, silvopastoral arrangement, chontaduro crops, undergrowth cocoa, five-year stubble, second-growth stubble and forest. Ten sampling units per cover were used, evaluating parameters of abundance, density and biomass in relation to some physical characteristics of the soil. The results showed significant differences between the total numbers of eggs and adults in the silvopastoral system vs. grasslands and in the semi-grazed grassland vs. ungrazed grassland and of adults in the chontaduro vs. cacao cover. While in natural forest the species are equally distributed, in the other cover crops there are different distribution patterns. Correlations were found between earthworm biomass, density and abundance with phosphorus and magnesium contents and soil bulk density.

"The results of the sampling carried out by Chamorro for the nine profiles, eight classes were recorded: Insecta. Arachnida. Diplopoda, Quilopoda. Symphila, Crustacea, Clitelata and Gasteropoda, belonging to the phylum: Artrhopoda. Annelida and Mollusca.

CONSIDERATIONS

a Soil contains at least a quarter to a third of all living organisms on the planet, but only about 1% of soil macro-organisms have been identified".

To achieve sustainability and resilience of the countryside, it is necessary to understand the functions performed by the soil, to improve management and, above all, to understand the role played by the biodiversity that exists there, giving it the relevance it deserves in terms of its function.

According to official sources such as the FAO, climate change and human activities - including poor management of agricultural practices - are leading to a loss of soil biodiversity and its functions at a current estimated rate of extinction between 100 and 1,000 times faster than would be considered natural.

It is important to know what has happened with climate change, how the habits of these organisms have changed and what the repercussions will be on soils and what consequences will be seen at the level of terrestrial ecosystems; this is a major task for the next generations of researchers.

BIBLIOGRAPHY

Brevik, E. C. 2014 The interdisciplinary nature of SOIL. Soil Journal Soil Journal 2014 The interdisciplinary nature of SOIL.

CANTIL, L. F. 2016. Present-day and fossil brood balls of necrophagous beetles. Thesis.

CHAMORRO, C. 1996. Avances del cono- cimiento Bioedafologico en Colombia. Revista Universidad Nacional de Colombia.

CHAMORRO, C.; Garcia 1998. Contribution to the knowledge of the temporal dynamics of the edaphofauna in a high Andean forest in the Monserrate region. Revista Universidad Nacional de Colombia.

CHAMORRO, C. 1982. Influencia del medio edáfico en la población de lombriz del suelo en la Sabana de Bogotá. Boletín del dpto. de Biología de la UN de Colombia.

CHAMORRO, C.; Amat 1987 Changes in the macrofaunal community in Vichada soils (Quartzipsamment) affected by burning CHAMORRO, C. 1989. Effect of land use on the edaphofaunistic composition of the soils surrounding the city of Bogota. Volume XIX of the SCCS journal Suelos ecuatoriales.

CHAMORRO C. LEON T. 1989. Ecological characterization of native earthworms (pheretima sp., eudovoscolex sp. and periscolex sp.) under different land uses (Guaviare, Colombia). Colombian Agronomy 1998 Volume XV No. 23 CHAMORRO C. et al. 1989. Estudio Bioeda- fologico Preliminar de La Isla Gorgona Acta biológica colombiana.

CHAMORRO C. MARTINEZ N., RO-
DRIGUEZ. 1992. Technological adaptation of
lumbering in the coffee-growing area of Alban Cundinamarca CHAMORRO, C. et al. 1995. Characterisation of the Edaphofauna and its dynamics in relation to land use. Aspectos ambientales para el ordenamiento territorial del Municipio de Mitú, departamento del Vaupés, Tomo III. IGAC.

CHAMORRO C. CAMERO E. 2021. Anthro-pofauna associated with high Andean forest soils of the Colombian Massif. Revista Entomotropica, Vol. 36.

CORTES, A. 1985. Soil and its diversity. Re- vista Universidad Jorge Tadeo Lozano. Bogotá.

FUENTES A., CHAMORRO C., LEON T.
1998. Colombian Agronomy Volume XV. No. 2.

GALVIS, H. VALENCIA, Z. CHAMORRO
C., CORTES, A. 1975. Edaphic effect of the army ant Atta laevigata in some soils of the integrated development centre Las Gaviotas in the Colombian Orinoquia. Thesis National University.

HALFFTER, G. 1997. Subsocial behavior in Scarabaeinae beetles. In: Choe J.C. and Cresspi B.J. (Eds.). The Evolution of Social Behaviour in In- sects and Arachnids. Cambridge University Press.

HUHTA, V. 2006. The role of soil fauna in ecosystems: A historical review. Pedobiologia 50 (2007) 489-495.

LAVELLE, P., Spain, A. (2001). Soil Ecology.
Kluwer Scientific Publications, Amsterdam, The Netherlands.

LAVELLE, P. 1996. Diversity of soil fauna and ecosystems function. Biology International. 33

LEON S. T. 2021. The main agroecological structure of agroecosystems. National University. Pag. 217 MONROY, F; Aira, M.; Domínguez J. 2011.
Facultad de Ciencias Exactas y Naturales, Uni- versidad de Buenos Aires NUÑEZ, B. 2015. Biodiversity, organic matter and soil structure. Doctoral thesis.
University of Burgos Spain.
PEDOBIOLOGIA (Journal of soil Ecolo- gy).2023.
PERNA, A. 2013. The art of building a nest.
Ed, Research and Science PINZON, A. 1976. Biological and edaphic aspects of the chiza, (dung beetle larvae) of some prairie soils of the Sabana de Bogotá.
PINZON, A. 1990. Characteristics of soils affected by the action of termites (paper).
PINZON, A. 2008. Soil physics. Zarate design. Pag. 205 PINZON, A. 2012. Soil science. Ed. Cargrap- fics. pag. 294.
PINZON, A. 2016. Soil physics II. Ed. ed.
Zarate design. Page 123.
PINZON, A; ROMERO M. 2006. Pedobiolog- ical action of soil fauna. World Congress of Soil Science. Philadelphia, USA, REDFERN, S. 2013. Excrements of lom- brices, a key to measure the climate of the past.
Geochimica et Cosmochimica RODRIGUEZ, T. 2023. Slugs and snails, from fertiliser and bioindicators of pollution to crop pests. University of Santiago de Compostela.
ROMERO, M. 1986. The earthworm as an indicator of soil contamination. Thesis Bogotá, Universidad de Bogotá Jorge Tadeo Lozano.
ROMERO, M. 1998. Transformation of organic waste through the rearing of soil lombrices.
ROMERO, M, 2002 Chemical characteristics of vermicompost made from water hyacinth and cattle manure. Revista Suelos Ecuatoriales.
Vol. 22.
ROMERO, M, 2002. Effect of vermicompost on soil fertility using lettuce as an indicator plant. Revista Suelos Ecuato- riales. Vol. 22.
ROYERO, S. 2019. Macrofauna edáfica y características físicas y químicas del suelo en áreas con diferentes sistemas de manejo en el departa- mento del Atlántico, Colombia. Thesis Universidad Nacional de Colombia Medellín, Colombia.
SCHULTZ, T. R. 2020. Fungus-Farming Ants. In Encyclopedia of Social Insects. Springer International Publishing.
SOTO, L. 1986. Protein content of the earthworm Andiodrilus bogotaensis. Thesis Bogotá, Universidad Nacional de Colombia.
TAPIA, C. 2019. Characterization of soil macrofauna in forest fragments in the municipality of Leticia, Colombian Amazon.
Revista Colombiana de Ciencia Animal, vol. 11, no. 1.
The role of soil fauna in ecosystems:mA historical review
VERSTEEGH, E. 2013. Earthworm droppings, a key to measuring past climate. BBC News World.

THE AUTHOR

Marfa Angela Pinzon Pinto

Agrologist, Bachelor in Biology and Master. Experience 25 years in the IGAC soil physics laboratory. Teaching and research at UJTL, UDCA, UNICUNDI. Direction of 10 thesis. Presentation of 5 investigations in world congresses and the same number in Latin American congresses, participation in the book "Amazonia y Agua" ed. of Unesco. Author of the following researches; Degradation index of the soils of the Amazonian Trapezoid. Compaction of soils by the action of cattle trampling in the Amazonian lomerio. Evaluation and behaviour of the physical properties in the municipality of Mitu. Change of the physical properties of the Andisols due to drying. Physical properties of soils in the Paramos of Chingaza, Neusa and Sumapaz. Evaluation of soil microstructure and porosity using computed tomography. Evaluation of fractal dimensions for application in soil porosity. Factors and mechanisms of formation of structural calcareous crusts in soils. More than 30 research papers published in the journal Suelos Ecuatoriales.

PEDOBIOLOGY

Maria Angela Pinzon Pinto

FSC
www.fsc.org

MIX
Papier aus verantwortungsvollen Quellen
Paper from responsible sources
FSC® C105338

Printed by Books on Demand GmbH, Norderstedt / Germany